CRAYOLA STEAM TEAMS

CREATIVITY, INNOVATION, AND TEAMWORK

Kevin Kurtz

Lerner Publications ◆ Minneapolis

To the STEAM teams on the R/V *JOIDES Resolution*

Copyright © 2021 by Lerner Publishing Group, Inc.

All rights reserved. International copyright secured. No part of this book may be reproduced, stored in a retrieval system, or transmitted in any form or by any means—electronic, mechanical, photocopying, recording, or otherwise—without the prior written permission of Lerner Publishing Group, Inc., except for the inclusion of brief quotations in an acknowledged review.

© 2021 Crayola, Easton, PA 18044-0431. Crayola Oval Logo, Crayola, Serpentine Design are registered trademarks of Crayola used under license.

Official Licensed Product
Lerner Publications Company
An imprint of Lerner Publishing Group, Inc.
241 First Avenue North
Minneapolis, MN 55401 USA

For reading levels and more information, look up this title at www.lernerbooks.com.

Main body text set in Mikado a Regular.
Typeface provided by HVD Fonts.

Editor: Andrea Nelson **Designer:** Laura Otto Rinne **Photo Editor:** Todd Strand
Lerner team: Sue Marquis

Library of Congress Cataloging-in-Publication Data

Names: Kurtz, Kevin, author.
Title: Crayola STEAM teams : creativity, innovation, and teamwork / Kevin Kurtz.
Description: Minneapolis : Lerner Publications, [2021] | Includes bibliographical references and index. | Audience: Ages 7–11 | Audience: Grades 2-3 | Summary: "Take a look at the many ways that teams use STEAM to solve problems. From developing smartphones to creating images of black holes, there's nothing teams can't do when they work together!"— Provided by publisher.
Identifiers: LCCN 2020014685 (print) | LCCN 2020014686 (ebook) | ISBN 9781728403229 (library binding) | ISBN 9781728423845 (paperback) | ISBN 9781728418582 (ebook)
Subjects: LCSH: Science projects—Juvenile literature. | Science—Experiments—Juvenile literature. | Group problem solving—Juvenile literature.
Classification: LCC Q182.3 .K87 2021 (print) | LCC Q182.3 (ebook) | DDC 507.8—dc23

LC record available at https://lccn.loc.gov/2020014685
LC ebook record available at https://lccn.loc.gov/2020014686

Manufactured in the United States of America
1-48293-48837-8/24/2020

Table of Contents

STEAM Teams Can Change the World   4

STEAM Teams Make Cities Smarter   8

STEAM Teams Uncover the Mysteries of the Brain   13

STEAM Teams See the Unseeable   18

STEAM Teams Help Save the World   23

My STEAM Dream Team   28
Glossary   29
Learn More   30
Index   31

STEAM Teams Can Change the World

It is hard to imagine our lives without smartphones. Over 2.5 billion people use them. But before 2007, hardly anyone knew what a smartphone was. Then Steve Jobs, the cofounder of Apple, first demonstrated the iPhone to the world.

Jobs sometimes gets credit for developing the first widely used smartphone. But he did not do it alone. He was the leader of a team of engineers and designers. Together, the team changed how people communicated and accessed information.

Apple cofounder Steve Jobs (*bottom*) may have presented Apple's newest innovations by himself onstage, but they were developed by hundreds of team members, including Scott Forstall (*top left*), Danika Cleary (*top center*), and Phil Schiller (*top right*).

Revolutionary ideas often come from the collaboration of teams. STEAM teams are groups of people with different knowledge and skills. They use science, technology, engineering, art, and mathematics (STEAM) to invent new things and solve problems.

The STEAM team that created the iPhone included software engineers like Ken Kocienda. He helped develop the iPhone's innovative touch screen keyboard. Artistic designers, including Apple's chief design officer Jony Ive, helped develop the iPhone's iconic sleek and minimalist look. They also designed the phone so that it felt nice to hold. Another designer, Imran Chaudhri, made sure the touch screen icons were attractive and easy to use.

Imran Chaudhri's clean and straightforward software revolutionized the way people used cell phones.

Yael Garten highlights Siri's popularity and showcases its newest features at an online developer conference in 2020.

Apple still has a STEAM team looking for ways to improve the iPhone. One member is Yael Garten. She is the director of data science for Siri, the iPhone's voice assistant. Garten uses mathematics to analyze the data on how people use Siri. She then determines the best ways to improve it.

STEAM teams work all around us. They solve problems and create new things. Sometimes these collaborations change the world.

STEAM Teams Make Cities Smarter

Cities tend to grow and change in unplanned ways. As a result, navigating streets can be confusing. Transportation and construction can be both expensive and bad for the environment.

Sidewalk Labs is a STEAM team working in Toronto, Canada. The team includes a variety of experts

TEAM STATS

Team name: Sidewalk Labs
Goal: Use data to improve the quality of life in cities and lessen their environmental impact
Location: Toronto, Canada
Key Team Members: Willa Ng, director of mobility, and Kristin Slavin, associate director, building innovations

Sidewalk Labs is owned by Alphabet, the same company that owns Google.

and innovators who are working to improve the quality of life for people who live in cities. They experiment with smart-city ideas in a neighborhood in Toronto. A smart city looks at data about how people use the city. It then installs technologies that solve problems the data shows exist.

One Sidewalk Labs team member is transportation engineer Willa Ng. Ng looks at how to use data to improve city traffic.

GROUNDBREAKING MOMENT

Sidewalk Labs developed the cross brace frame, an outer framework of wood beams. The frame is strong enough to allow all-wood buildings to be taller than ever before, up to thirty-five stories.

She is working on a new feature that could be added to navigation apps. The redesigned apps would tell drivers how much each trip costs them in gas, tolls, car maintenance, and more. If drivers know the true cost of each trip, they may be more likely to use cheaper transportation options such as bicycles, trains, or buses. Having fewer drivers on the road would reduce city traffic and greenhouse gas pollution. It would also reduce the amount of time people spend stuck in traffic jams.

Kristin Slavin is a Sidewalk Labs architect focused on building a more environmentally friendly city by developing innovative skyscrapers made out of wood. She is helping to develop a high-rise neighborhood in Toronto that would use these wooden buildings. New kinds of wooden building materials such as cross-laminated timber are fireproof and as strong as steel. Constructing wooden buildings produces far fewer greenhouse gases than constructing buildings made of concrete and steel. But wood skyscrapers are much lighter than traditional skyscrapers. They are more

Sidewalk Labs spent several years planning a smart neighborhood in Toronto that included driverless cars, heated sidewalks, and many data sensors.

vulnerable to forces of wind and gravity. Slavin and the rest of the Sidewalk Labs team had to figure out how to make their wooden skyscrapers more stable.

This STEAM team uses its Toronto neighborhood as a lab for their smart-city ideas. The ideas that work well can then be shared with cities around the world to make them more efficient, sustainable, and livable.

STEAM Teams Uncover the Mysteries of the Brain

Human brains are so complex that scientists are still learning about how they work. Technology is helping them learn more. Magnetic resonance imaging (MRI) machines use a magnetic field to create images of the inside of a human body. MRIs are giant, tube-shaped magnets.

TEAM STATS

Team name: National Institute on Deafness and Other Communication Disorders

Goal: To understand what happens in the brain during creative activity

Location: Bethesda, Maryland

Key Team Members: Siyuan Liu, neuroscientist; Katherine Swett Aboud, neuroscientist; and Michael Eagle, a.k.a. Open Mike Eagle, freestyle rapper

The opening in an MRI machine is just big enough for someone to lie in it. When the machine runs, a magnetic field passes harmlessly through the person to create images. Neuroscientists use MRI images to see which parts of the brain are active while a person does different things.

One question neuroscientists have is, What happens in the brain when someone creates a painting, a piece of music, or other forms of art? This question has been difficult to study, in part because you can't bring a musical instrument or art supplies into an MRI tube. Scientists needed another artistic activity—one that could fit in an MRI.

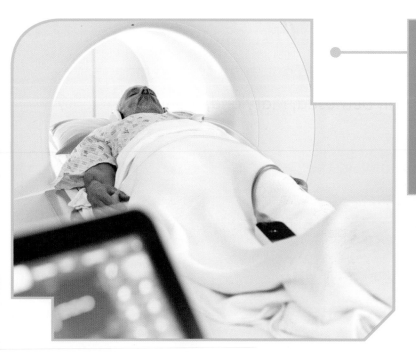

In the United States, millions of MRI scans are done every year as part of health care and research.

The National Institute for Deafness and Other Communication Disorders in Bethesda, Maryland, helped fund the new study.

A STEAM team in Bethesda, Maryland, found one activity they could study. Siyuan Liu, Katherine Swett Aboud, and other neuroscientists worked with twelve freestyle rappers, including Open Mike Eagle. Freestyle rappers don't memorize lyrics ahead of time. They invent rhymes as they go. Rapping can happen in an MRI tube. Rappers just use their mouth and their brain.

Rapper Open Mike Eagle is credited as one of the authors of the new study.

When the rappers entered the MRI, the neurologists played beats for them. The rappers then improvised rhyming lyrics. The rappers also performed rhymes that they had memorized beforehand.

The MRI took images of the rappers' brain activity. The scientists compared the rappers' brain activity during the improvised raps with their brain activity during the memorized raps. Open Mike Eagle helped evaluate each

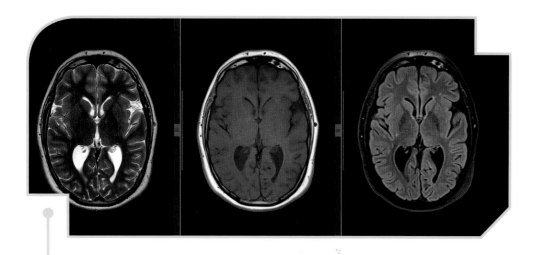

Researchers analyzed MRI scans like these to understand which parts of the brain are used for artistic improvisation.

improvised rhyme to determine which were the most creative. The scientists compared the brain activity during the more creative raps with the activity during less creative raps. They found that the analytical parts of the brain mostly shut down during improvisation, but that other parts of the brain involved in creative problem-solving became more active. This STEAM team gained a better understanding of what happens in the brain when someone is being artistically creative. But the researchers believe this same type of brain activity could happen in other fields where spontaneous creative thinking is necessary—including science.

STEAM Teams See the Unseeable

Black holes are the most massive objects in the universe. They have so much gravity that they suck even light into them. It seemed impossible for people to ever see one.

This assumption proved wrong. In 2019, a STEAM team project named Event Horizon Telescope took the first photo of a black hole. The image showed the silhouette of a supermassive black hole in

TEAM STATS

Team name: Event Horizon Telescope

Goal: Capture a visual image of a black hole

Key Team Members: Feryal Özel, astrophysicist; Jonathan Weintroub, engineer; Katie Bouman, computer scientist; and Katie Peek, artist

the Messier 87 galaxy. The black hole stood out against the bright backdrop of stars and light it was devouring.

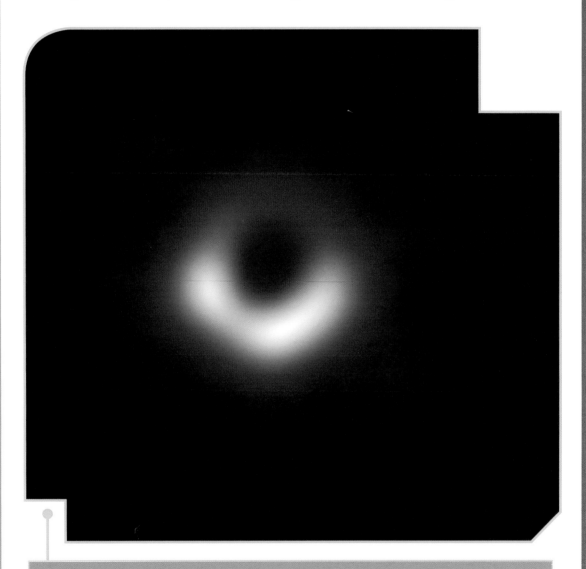

More than two hundred scientists across the globe worked together to create the very first image of a black hole.

Messier 87 is 55 million light-years from us. No telescope on Earth has the resolution to see an object that far away. But the Event Horizon Telescope STEAM team had an idea. What if they made a telescope that was as big as Earth?

The STEAM team relied on scientists and engineers to make the network of telescopes work. Engineers like Jonathan Weintroub worked on the technology to make sure the telescopes were exactly in sync. Computer scientists like Katie Bouman created algorithms that organized the

GROUNDBREAKING MOMENT

The team decided eight telescopes around the world would all focus on the black hole at the same time. Working together, they created a telescope "lens" the size of planet Earth.

telescopes' data to create an image. Astrophysicists like Feryal Özel analyzed the image and then collected data to learn more about black holes.

Artists were also part of the team. The main goal of science is to increase human knowledge. This requires communication. Scientists often use art to help explain their findings in a way that is easily understood. The Event Horizon Telescope team worked with artist Katie Peek. She created a comic strip to explain to people who are not

Katie Bouman and her team developed multiple algorithms to analyze the telescope data and ensure the image accurately represented the black hole.

The Event Horizon Telescope includes radio telescopes from around the globe, including some from the Atacama Large Millimeter/submillimeter Array (ALMA) in northern Chile.

scientists how electromagnetic waves travel from a black hole to the Event Horizon Telescope.

Working together, the Event Horizon Telescope STEAM team helped scientists learn more about the universe and the roles black holes play in it.

STEAM Teams Help Save the World

The world's oceans have been filling with plastic garbage. Some of the litter is large, but most of it is microplastics. Microplastics are plastic garbage such as straws and grocery bags that have broken down in the sun, wind, and waves. The tiny plastic

TEAM STATS

Team name: Monterey Bay Aquarium
Goal: Create art that will educate the public about ocean pollution
Location: Monterey Bay and San Francisco, California
Key Team Members: Joel Dean Stockdill, artist; Yustina Salnikova, artist; and Rbhu, engineering firm

Approximately 9 million tons (8 million t) of plastics enter the ocean each year.

pieces float in the water for many years. This marine debris is dangerous to marine life. Ocean animals accidentally eat the microplastics or get trapped in the larger garbage.

The Monterey Bay Aquarium in California wanted to make more people aware of this serious problem. The aquarium put together a STEAM team to help. The artists Joel Dean Stockdill and Yustina Salnikova worked with the aquarium, engineers, and other artists. They created a life-size blue whale made of plastic garbage. Blue whales are the largest animal on Earth. They can weigh about 300,000 pounds (136,078 kg). That is also how much plastic garbage ends up in the ocean every nine minutes. When completed, the sculpture was 82 feet (25 m) long and weighed about 2 tons (1.8 t).

The Monterey Bay Aquarium in California has been championing ocean conservation for over thirty-five years.

The STEAM team needed engineers to design a framework that could support the giant sculpture.

The finished project was unveiled in 2018. After spending time in San Francisco, the whale traveled to various states. It has helped to educate many people about the serious problem of marine debris.

The blue whale art piece was unveiled at Crissy Field in San Francisco.

Jared Chen, one of the team members, reinforces panels inside the whale sculpture.

People can accomplish a lot when they work together. Some day you might start a STEAM team to create new ideas and solutions. Your team's innovations could change the world.

MY STEAM DREAM TEAM

You can form your own STEAM team with your friends to solve problems and create and discover new things. Follow these steps to put together your own STEAM dream team:

- Bring together a group of friends with different knowledge, talents, and skills.

- Identify a problem the group wants to solve or a question they want to answer. The issue could be big, like trying to stop climate change, or smaller, like creating habitats for local wildlife or reducing noise pollution in your neighborhood.

- Have everyone do research to learn about the issue.

- Brainstorm ideas about what you want to do.

- Create models of the team's best ideas.

- Test your models to see what works and what needs to be changed.

- Once you have a model that works, share your idea with others!

GLOSSARY

algorithm: a step-by-step method for solving a problem

astrophysicist: someone who studies the behavior of objects and events in space

data: facts and information that are collected to study and learn from

electromagnetic waves: different wavelengths of energy that travel with light

engineer: a person who uses math and science to figure out the best way to build things

freestyle: a type of rap where artists make up lyrics on the spot

gravity: an invisible force that attracts objects to one another

greenhouse gas: a type of gas in the atmosphere that can trap heat

improvise: to create and perform without preparation

neuroscientist: a person who studies the brain and nervous system

resolution: a camera's or telescope's ability to bring an image into focus

LEARN MORE

Engineering, Go For It (eGFI)
http://www.egfi-k12.org

Exploratorium: Explore, Play, Discover
https://www.exploratorium.edu/explore

Miller, Derek L. *Group Planning, Creating, and Testing: Programming Together*. New York: Cavendish Square, 2018.

Turner, Matt. *Genius Engineering Inventions: From the Plow to 3D Printing*. Minneapolis: Hungry Tomato, 2018.

Wainewright, Max. *Design, Animate, and Create with Computer Graphics*. Irvine, CA: QEB, 2017.

INDEX

Apple, 4–7

black hole, 18–22

comic strip, 21
creativity, 17

iPhone, 4–7

magnetic resonance imaging (MRI) machine, 13–17
microplastics, 23–24
Monterrey Bay Aquarium, 23–25

pollution, 10, 23

rapping, 15–17

Sidewalk Labs, 8–12
skyscraper, 11–12

telescope, 18, 20–22
traffic, 9

whale, 25–27

PHOTO ACKNOWLEDGMENTS

Image credits: AP Photo/Jeff Chiu, p. 5; Rajat Bhardwaj/flickr (CC BY-S2.0), p. 6; Independent Picture Service, p. 7; STR/AFP/Getty Images, pp. 9, 11; Juice Flair/Shutterstock.com, p. 14; bakdc/Shutterstock.com, p. 15; Hutton Supancic/Getty Images, p. 16; springsky/Shutterstock.com, p. 17; Event Horizon Telescope collaboration et al., p. 19; National Science Foundation, p. 21; Iztok Bončina/ESO (CC BY 4.0), p. 22; Shane Gross/Shutterstock.com, p. 24; hotocritical/Shutterstock.com, p. 25; AP Photo/Eric Risberg, pp. 26, 27. Robert Kneschke/Alamy Stock Photo, p.28. Design elements: Batshevs/Shutterstock.com; iD_studio/Shutterstock.com.

Cover: Batshevs/Shutterstock.com; ID_studio/Shutterstock.com (background).

BROOKLINE VILLAGE
PUBLIC LIBRARY OF BROOKLINE
361 WASHINGTON STREET
BROOKLINE, MA 02445